D0534119

Focus on Alcohol

Focus on Alcohol
A Drug-Alert Book

Catherine O'Neill
Illustrated by David Neuhaus

Troll Associates

Printed in the United States of America

10 9 8 7 6 5 4 3 2 1

Library of Congress Cataloging in Publication Data

O'Neill, Catherine
Focus on Alcohol
Illustrated by David Neuhaus

(A Drug-Alert Book)
Includes bibliographical references.
Summary: Discusses the history, use, and
dangers of alcohol, the problems of alcoholism,
and coping with the pressures to drink.
1. Alcoholism—United States—Juvenile literature.
[1. Alcohol. 2. Alcoholism.]
I. Neuhaus, David, ill. II. Title.
III. Series: The Drug-Alert Series.
HV5066.054   1990
362.29′2′0973—dc20   89-20410   CIP

Table of Contents

Introduction

"Baby Saved by Miracle Drug!" "Drug Bust at Local School!" Headlines like these are often side by side in your newspaper, or you may hear them on the evening news. This is confusing. If drugs save lives, why are people arrested for having and selling them?

The word "drug" is part of the confusion. It is a word with many meanings. The drug that saves a baby's life is also called a medicine. The illegal drugs found at the local school have many names—names like pot, speed, and crack. But one name for all of these illegal drugs is dope.

Some medicines you can buy at your local drugstore or grocery store, and there are other medicines only a doctor can get for you. But whether you buy them yourself or need a doctor to order them for you, medicines are made to get you healthy when you are sick.

Dope is not for sale in any store. You can't get it from a doctor. Dope is bought from someone called a "dealer" or a "pusher" because using, buying, or selling dope is against the law. That doesn't stop some people from using dope. They say they do it to change the way they feel. Often, that means they are trying to run away from their problems. But when the dope wears off, the problems are still there—and they are often worse than before.

There are three drugs we see so often that we sometimes forget they really are drugs. These are alcohol, nicotine, and caffeine. Alcohol is in beer, wine, and liquor. Nicotine is found in cigarettes, cigars, pipe tobacco, and other tobacco products. Caffeine is in coffee, tea, soft drinks, and chocolate. These three drugs are legal. They are sold in stores. But that doesn't mean they are always safe to use. Alcohol and nicotine are such strong drugs that only adults are allowed to buy and use them. And most parents try to keep their children from having too much caffeine.

Marijuana, cocaine, alcohol, nicotine, caffeine, medicines: these are all drugs. All drugs are alike because they change the way our bodies and minds work. But different drugs cause different changes. Some help, and some harm. And when they aren't used properly, even helpful drugs can harm us.

Figuring all this out is not easy. That's why The Drug-Alert Books were written: so you will know why certain drugs are used, how they affect people, why they are dangerous, and what laws there are to control them.

Knowing about drugs is important. It is important to you and to all the people who care about you.

David Friedman, Ph.D.
Consulting Editor

Dr. David Friedman is Associate Professor of Physiology and Pharmacology and Assistant Dean of Research Development at the Bowman Gray School of Medicine, Wake Forest University.

The Alcohol Problem

It's the first day of spring. You feel a breeze pass through the hot and crowded cafeteria. After lunch, you and your friends hurry out to the playground. It's nice to be outside and feel the first warm sunshine of the year. You sit down for a few minutes on the playground bench.

Soon your best friend sits down next to you. She whispers important news to you. Right after school, she and some other kids are going to get some beer. And she asks you to come along. "Come on," she says, "we're going to have a party!"

At first, you are too surprised to answer. So many questions are running through your head. How are they going to get the beer? Where are they going to drink it? What will they say to their parents? But your friend is very excited about her plans. And, once again, she asks you to come.

"What should I say?" you ask yourself. "What should I do?"

It takes a while to get used to the idea that your friend wants to drink beer. You thought you knew her so well. You never knew she drank beer! You think of all the times your parents told you to say "No" if you were asked to use drugs. But now your best friend is asking you to say "Yes." Your best friend!

You don't really want to. If you do, you will have to lie to your parents about why you got home from school late. Even if they believe you, it feels bad to lie to them. And what if you get caught? You don't even want to think about that! Anyway, you don't really like the way beer smells. Why would you want to taste it?

Your best friend is waiting for your answer. If you say "No," she'll be disappointed. You can just tell. She might even be mad.

And you know what the other kids will say about you. You just know.

Your mind is racing.

Should you say "Yes"? Should you say "No"?

"What is so bad about a little beer?" you might ask yourself. You see alcohol everywhere. Signs in stores point the way to "Cold Beer!" Signs in restaurant windows read "Fine Food and Drink." On the radio you hear that beer "tastes great" or "goes down smooth." On TV you see commercials that make you think everyone likes alcohol: athletes drink foaming mugs of beer after a game, and families lift their glasses of wine around the dining room table.

Maybe your parents have a drink together before dinner. Or maybe your grandparents have a drink when dinner is done. Away from home, you see people drinking beer at ball games or sipping champagne at wedding receptions.

If alcohol is everywhere, can it really be so terrible? Sitting with your best friend in the warm sunshine, you ask yourself, "Why shouldn't I just say 'Yes'?"

The answer is simple: alcohol is a dangerous drug. It changes the way our bodies and minds work. It changes the way we think and feel and behave. Many adults who drink alcohol are hurt by it. And like all drugs, alcohol is especially dangerous for young people. That's why it is illegal for them to use it.

There is a special reason to be concerned about children and alcohol. Alcohol is a gateway drug. This means that many people who start using alcohol early in life end up with drinking problems. They may go on to use other drugs, too. Using alcohol helped to "open the gate" to drug problems.

You still have not told your best friend whether you will go drinking with her after school. If you know more about alcohol, it will be easier for you to say "No" to your friend. So, before you answer her, let's find out about this drug. In this book, we will find out why people use alcohol and what happens to them when they do. We will find out what alcohol is and what it does to the body and the brain. We will find

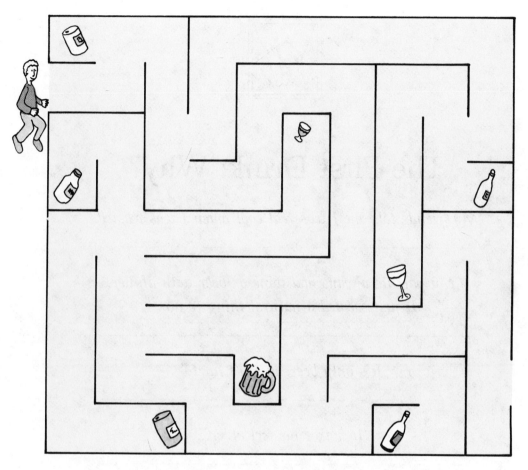

out how alcohol was used long ago and how it is used today. And, finally, we will find out what kind of decisions you will face as you live and grow up in a world where alcohol is all around you.

Your decisions may not be easy ones, but they have to be made. This book can help you make the right ones.

The First Drink: Why?

"My friends told me I was real cool when I was wasted."

*"My parents always had wine or beer with dinner.
It just seemed a natural thing to do."*

"It's only beer. It can't hurt."

"It makes me feel older."

These are the words of real kids. They are talking about themselves and their drinking. They have a drinking problem, and they are not alone. In this country, over 3 million young people are in trouble with alcohol. One out of every five teenagers has a drinking problem.

Why do kids use alcohol? There are many reasons. Here are some of them:

• Peer pressure

Your peers are the people you spend time with every day. They are the people you go to school with, play sports with, and hang out with. Peer pressure is the feeling that you have to do the things your peers want you to do. Feeling this kind of pressure is a normal part of growing up. All of us want to be like others. All of us want others to like us.

Peer pressure might mean listening to a certain kind of music, wearing a popular style of clothes, or joining the "in" club. But, sometimes, peer pressure means doing things like skipping school or drinking beer with your friends. One of the reasons kids most often give for taking their first drink is to "fit in with my friends."

• Parent pressure

It's only natural for kids to want to be like their parents. They learn by listening to what their parents say. They learn even more by watching what their parents do.

Many young people learn to drink at home. They see their parents drinking. If parents show their children that drinking is okay, young people are more likely to try it, too.

• Pressure from advertising

On television, in magazines, and on the radio, the people who sell alcohol make drinking look like fun. They sell alcohol by making people think that it's "cool" to drink, by making people think that drinking helps them have a good time, and by making people think that everyone drinks.

And they sell alcohol by *not* telling people that drinking is dangerous. Advertisers want people to think that a beer is harmless or that a wine cooler is just like a soft drink.

• Pressure to grow up

A lot of kids want to grow up in a hurry. They think drinking makes them more adult. Other kids may already lead almost grown-up lives: earning money to help their families or taking care of their younger brothers and sisters while their parents work. These young people may think they are old enough to drink. Or they may drink to escape the pressure to grow up fast.

All of these pressures are a normal part of growing up. Growing up can be a hard time. Kids often feel alone and confused. But drinking doesn't help with the pressures of growing up: it only makes them worse.

That's what kids who drink find out the hard way. Once young people start to drink, they find that drinking becomes more and more a part of their lives. They have friends who drink. They go to parties where people drink. And before they know it, they find that drinking is more important to them than school or friends or family.

These unlucky kids find themselves trapped. They find themselves trapped in a world inside a bottle.

What Is Alcohol?

Alcohol is a clear, colorless liquid. It is found in three kinds of drinks: beer, wine, and liquor. Whiskey, rum, gin, and vodka are examples of liquor.

The kind of alcohol people drink is called ethyl alcohol or ethanol. There are other kinds of alcohol that are not used for drinking. One of them you may know as rubbing alcohol. It's what the doctor rubs on your skin before you get a shot. It is also called isopropyl alcohol. Isopropyl alcohol is very poisonous to drink. In this book, the word "alcohol" will mean ethyl alcohol, the kind found in beer, wine, and liquor.

Some alcoholic drinks are stronger, or contain more ethyl alcohol, than others. Beer has the least amount of alcohol in it. Wine has a higher amount—more than twice as much as beer. And liquor has the highest amount of all—more than eight times as much as beer does.

Look at this picture of three glasses: a drink of beer, a drink of wine, and a drink of whiskey. Even though they hold different amounts of beer, wine, or whiskey, the amount of alcohol in each drink is the same.

In this book, a "drink" means 12 ounces of beer, 5 ounces of wine, or 1-1/2 ounces of liquor.

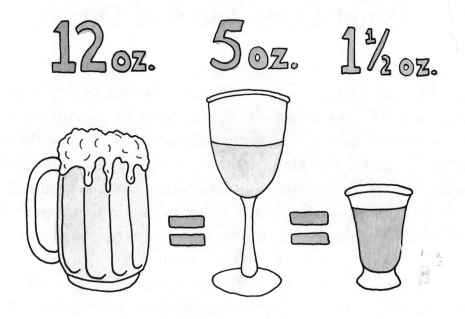

Because beer and wine are not as strong as liquor, some people think they are safer to drink. But one kind of alcoholic drink is *not* safer than another. The alcohol in each kind of drink is the same drug.

What Does Alcohol Do?

What happens to someone who drinks alcohol?

Just like a peanut butter and jelly sandwich or a glass of milk, alcohol goes into the stomach. Most food or drink is broken down or digested in the stomach and the intestines before it enters the bloodstream. This takes time, usually several hours. But alcohol is not like other food and drink: it doesn't need to be digested. Alcohol goes straight from the stomach or intestines into the bloodstream. The bloodstream carries it to every part of the body, including the brain.

Alcohol does not stay in the body. The body gets rid of it in several ways: sweating, urinating, and even breathing all help the body use up alcohol. But almost all of the work has to be done by an organ in the body called the liver. The liver changes alcohol into water and a harmless gas called carbon dioxide. It takes the liver one hour to get rid of the alcohol in one drink. The liver just can't work any faster than that.

But people often have more than one drink an hour. They drink alcohol faster than the liver can get rid of it. So the alcohol builds up in their blood. And it builds up in the brain, too. That's when we say a person is drunk or intoxicated.

What happens when alcohol reaches the brain?

Alcohol changes the way the brain works. You can understand this better if you think of your brain as a message center. Your body sends messages to the brain so you know about the world around you. When you see or hear something, your body is sending the brain messages and telling it what is going on. Like: "Hey, there's a speeding train heading this way!" Your brain, in turn, sends out messages to control your body. When you move, for example, your brain is sending the body messages and telling it what to do. Like: "Don't just stand there! Get out of the way!"

Your brain also sends messages to itself. Information has to get to different parts of the brain so we can examine it, remember it, and use it to guide our actions. That's when thinking occurs: when the different parts of the brain send messages to each other.

Stop! Right now, what are you doing?

You're reading. Your brain tells you where on the page to look and how to understand the words. Is reading all that you are doing right now? You're also holding this book. Your brain tells your arms and hands how to work. But is that all you are doing? No. You are also breathing, and your heart is beating, and your body is changing the food you eat into the energy you need to play and think and dream. Your brain tells your lungs and heart and stomach what to do. Is that all? Right now, you're also thinking about your brain. And even though you can't see it or feel it, your brain is telling your body to grow. Your brain sends out billions and billions of messages every day.

The brain has a big job to do. It can do that job well when people are sober: that's when there is no alcohol in the body. But when alcohol reaches the brain, it depresses or slows down the way the brain works. (Alcohol is a kind of drug called a depressant.) It disrupts some of the messages the brain sends to itself and to the body. That means some messages get lost, and others are slow or unclear. This makes it hard to do all the things people need to do: make decisions, pay attention, see things well, speak clearly, walk straight, even stand up.

A person who has been drinking would have a hard time reading this book. It would be hard:

-- to see the page clearly
-- to hold the book straight
-- to pay attention to the words
-- to understand or remember the book
-- to talk to other people about the book

Different people behave differently when they are drunk. Some people may have trouble walking and talking. People may say or do things they wouldn't say or do when they are sober. One person may get sad for no reason, another person may get loud and rowdy, and still another may get angry and violent. If they keep on drinking, people may get sick to their stomachs or even pass out. After people stop drinking, they don't feel very well. They may have a bad headache or an upset stomach. They may feel tired and miserable. This sick feeling is called a hangover.

People who drink will get over the sick feeling of a hangover. But if they continue to drink, they will face a problem much worse than a headache or upset stomach. Alcohol is a very addictive drug. It changes the way the brain works so that people begin to need the drug to feel normal.

The messages to and from the brain tell us how to lead a healthy life. If we need rest or sleep, the brain tells us we are tired. If we need food or drink, the brain tells us we are hungry or thirsty. But alcohol changes the way the nerve cells of the brain work. They begin to send messages that say the body needs alcohol, just as they send messages that tell us when we need food and sleep. That's why people who are addicted to alcohol feel that they *must* have alcohol. They don't just *want* to drink: they feel that they *need* to drink.

How much, then, is it safe for a person to drink?

No amount of alcohol is safe for young people to drink. Even one drink will slow down the brain and change the way a person thinks and feels and behaves. Just one drink: one glass of beer, or one wine cooler, or one shot of whiskey.

Alcohol: True or False?

1. Coffee will help a drunk person get sober.

>**False.** Coffee will help to keep a drunk person awake, but it will not stop someone from being drunk. Neither will a cold shower or fresh air. The only thing that helps is time, time for the liver to get rid of the alcohol.

2. Mixing different kinds of drinks makes people more drunk.

>**False.** The alcohol in different drinks is the same, and it is the alcohol that gets people drunk. Mixing different kinds of drinks will not get a person more drunk, and drinking only beer, or only wine, or only liquor won't keep a person from getting drunk.

3. Alcohol makes people forgetful.

>**True.** Alcohol makes it harder for a person to remember things. In fact, some people who have been drinking for many years may not be able to remember anything new at all.

4. A person could die from drinking too much alcohol.

> **True.** Drinking a lot of alcohol too fast could stop the brain from sending messages to a person's body. And without the brain telling the lungs to keep breathing and the heart to keep pumping, a person would die.

5. Alcohol is a good "pick-me-up." It gives people energy.

> **False.** Alcohol is a depressant. It slows down the way the brain works. A small amount of alcohol might make some people feel relaxed. But it will never give anyone energy.

6. Eating will keep you from getting drunk.

> **False**. Food may slow down the effects of alcohol, but it does not stop alcohol from entering the bloodstream. So, although it may take a little longer, people can and will get drunk on a full stomach.

7. One drink can't hurt anyone.

> **False.** Even one drink changes the way people think, feel, and behave. There is no safe amount of alcohol.

Alcohol: A History

No one is sure how people first learned to make alcohol. But we do know that there have always been people who used alcohol. The ancient Greeks and Romans drank it. The lords and serfs in the Middle Ages drank it. And when the Puritans came to America in 1620, they brought the practice of drinking alcohol with them.

The Puritans made their own alcoholic drinks. They used different fruits, like apples, grapes, and berries. They also used vegetables, like carrots, celery, and spinach. For those who did not make their own beer or wine, there were many public drinking places called tippling houses. And inns and taverns served alcohol to the weary travelers along the new roads.

In those early days, clean drinking water was very hard to find, so children as well as adults drank alcohol with their meals. It was even served at breakfast!

But alcohol was more than just another drink. It was a way to celebrate important events: when a new house was built in the spring or when the fall harvest was brought into the barn. Alcohol was served on special occasions such as weddings and even funerals. It was served at town meetings and on election days. It was used as a medicine to ease pain, or to bring down a fever, or to soothe an upset stomach.

Although drinking was accepted by nearly everyone, drinking *too much* was not. "Drink is a good creature of God," wrote one leader, "but the drunkard is from the Devil." People were punished for getting drunk. In 1633, for example, a man named John Holmes was ordered to "sit in the stocks" for being drunk.

As America grew, so did the amount of alcohol people drank. In the 1700s and 1800s, Americans were drinking more than twice as much as they drink today. They were known as "a nation of drunkards." Why did Americans drink so much then? Here are some of the reasons:

- People were moving away from the farms and into the bustling and crowded cities. There, they lived far from the people they loved and the life they knew. They drank to try to forget the loneliness of life in the new cities.

- People were coming to America from many other countries in search of better lives. But their lives here were often hard and poor ones. They drank to try to forget the pain of a new life in a strange country.

- People were settling the frontier towns of the West. Cowboys, lumberjacks, trappers, and miners lived rugged lives. They drank to try to forget the hardships of life on the frontier.

What's in a Name?

In the 1700s and 1800s, Americans were called "a nation of drunkards." There were many people who drank too much—and many names to describe them. Benjamin Franklin, the American newspaperman and diplomat, collected more than 230 different names for a drunkard. Here are just a few of them:

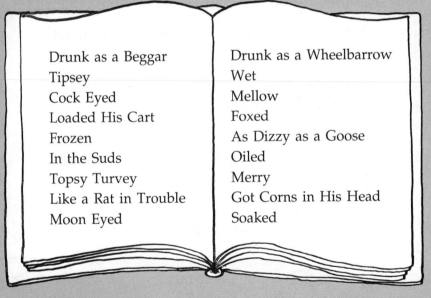

Drunk as a Beggar
Tipsey
Cock Eyed
Loaded His Cart
Frozen
In the Suds
Topsy Turvey
Like a Rat in Trouble
Moon Eyed

Drunk as a Wheelbarrow
Wet
Mellow
Foxed
As Dizzy as a Goose
Oiled
Merry
Got Corns in His Head
Soaked

from Benjamin Franklin's "Drinker's Dictionary,"
Pennsylvania Gazette, January 6, 1737

From the cities of the East to the barrooms of the West, people were drinking. But not everyone approved of alcohol. There were men and women who thought that people should temper, or slow down, the amount of their drinking. They formed the American temperance movement. One temperance leader was Benjamin Rush, who lived from 1745 to 1813. Rush was one of the first people to realize that drinking too much was an illness: it did not mean that the drinker was a bad person. In 1833, 20 years after Rush died and 200 years after John Holmes was put in the stocks, the temperance leaders started a new organization, the American Temperance Union.

Another leader of the temperance movement was Lyman Beecher. Like Rush, he wanted to temper America's drinking habit. But, unlike Rush, Beecher wanted liquor to be made illegal. Other people, called prohibitionists, believed no one should be allowed to drink any alcohol, including beer and wine. One of the most important of these prohibitionists was Frances Willard. In 1874, she formed a new temperance group, the Women's Christian Temperance Union, to teach people about the dangers of drinking.

Some prohibitionists didn't think *teaching* people about alcohol was enough. They wanted to *show* people the dangers of drinking. That's just what Carrie Nation tried to do. She attacked barrooms with a hatchet, chopping open barrels of beer and breaking bottles of whiskey. "Smash! Smash!" she would cry as her ax flew. Carrie Nation attacked saloons from Kansas to New York.

In the early 1900s, the prohibitionists formed their own political parties. They worked hard to convince others that drinking alcohol was wrong, and they were quite successful. In 1920, the Eighteenth Amendment to the Constitution made it illegal to make, sell, or ship alcohol—everywhere in the United States. This period of time is called Prohibition.

But there was plenty of illegal or "bootleg" alcohol to drink. It was smuggled into the United States from Canada and Mexico. It was made in illegal factories called stills. It was also made in some homes. Bootleg alcohol was served in private clubs called speakeasies. It was even served in the White House.

Soon, gangsters known as bootleggers were making and selling illegal alcohol. To gang leaders like Al Capone, alcohol meant money—and lots of it. In many cities, mobsters fought the police and each other to control the production and sale of bootleg liquor.

It was not long before most people, even if they were against drinking, wanted to see the prohibition law changed. On December 5, 1933, only 13 years after it had been voted *in*, the Eighteenth Amendment was voted *out*. Making, selling, and shipping alcohol became legal again.

From then on, each state set its own rules about drinking. From state to state, there were many different laws about:

- who could make alcohol

- how old a person must be to drink alcohol

- what kinds of stores could sell alcohol

- when people could buy alcohol

- where people were allowed to drink alcohol

One law that caused a great deal of debate was the legal drinking age. Some states ruled that the drinking age should be 21, others set the drinking age at 19, and still others at 18. But people did agree that alcohol was hurting too many teenagers. Too many teenagers had drinking problems. Too many teenagers were drinking and driving. And too many teenagers were being hurt and killed (and hurting and killing others) in thousands and thousands of car crashes.

People across the nation wanted to keep teenagers from drinking and driving. Many felt that teenagers would be safer if the legal drinking age were 21. By the 1970s and early 1980s, the states began to raise the legal drinking age. Today, the different states still have different laws about making and selling alcohol. But one rule is the same everywhere: you may not drink alcohol unless you are 21 years old.

Carrie Nation: Portrait of a Booze Buster

Carrie Amelia Moore was born in 1846. When she was 19, she married Dr. Charles Gloyd. Her husband drank a great deal, and his drinking made Carrie's life a very unhappy one. Carrie never forgot the sad years with him, even after he died and she was remarried to David Nation. She spent the rest of her life telling others of the dangers of alcohol.

In the 1880s, Carrie Nation joined the Women's Christian Temperance Union in Kansas. In 1900, she began her crusade against alcohol, visiting saloons with her famous hatchet. Carrie traveled the entire country, carrying out her "hatchetations," as she called them. Six feet tall and over 175 pounds, Carrie Nation was a hard person to stop. But she was stopped, and often: she was arrested over 30 times.

Carrie A. Nation once said her name meant "carry a nation to temperance." She died in 1911, eight years before the beginning of Prohibition.

Al Capone: Portrait of a Bootlegger

Alphonse Capone was born in Italy in 1899 and grew up in Brooklyn, New York. He lived in a tough neighborhood, and he was as tough as the streets where he grew up. In one fight, he was slashed with a razor blade on his left cheek. He was left with a scar from his ear to his lip. From then on, he was known as Scarface.

Al Capone soon entered the underworld of crime. He moved to Chicago and worked for a gangster who sold bootleg alcohol. In the 1920s, Capone became the boss of the entire Chicago crime world. His rule was very bloody and violent. On one especially brutal day, Capone's gunmen, dressed as policemen, gunned down seven rival gang members. That day, February 14, 1929, is known as the St. Valentine's Day Massacre.

In 1931, Al Capone was sent to prison. He was a sick man when he was let out in 1939. Al Capone, the king of bootleg liquor, died in 1947.

Alcohol Today

Did you see the movie *ET*? In one scene ET drinks beer for the first time. He drinks one beer, then another, and then another, until he is drunk. It makes everyone laugh to see this funny little creature bumping into walls and falling over himself. But is it really so funny to get drunk? Is drinking really something to laugh about?

Not when you look at the facts on drinking alcohol in America today. Here are some of them:

- Each year over 100,000 people die from sicknesses caused by alcohol.

- Each year 500,000 people go to the hospital with sicknesses caused by alcohol.

- Each year over 4,000 babies are born with serious illnesses because their mothers drank alcohol while they were pregnant.

One of the worst problems caused by alcohol happens when someone drinks and drives:

- Drinking and driving kills over 3,600 teenagers each year. That's 10 teenagers killed every day.

- Over 23,000 adults are killed each year in drinking and driving car crashes. That's one adult killed every 20 minutes.

- Because of drinking, over 85,000 teenagers are hurt each year in car crashes.

To you, these may just be numbers. But behind each of these numbers is a real person: a person who has friends and family, a person who was hurt or killed by alcohol. And it happens every day. It happened to Cari Lightner when she was killed by a driver who had been drinking.

On May 3, 1980, Cari was walking to a school carnival. What should have been a day of fun with her friends was the last day of her life. She was hit by a car driven by a man who had been drinking. The car sped away, and Cari was dead. She was 13 years old.

Cari's mother learned that the driver who killed Cari had been arrested for drunk driving in 1977, 1978, and 1979. And he had been arrested again for drunk driving just two days before Cari's death! Yet he was still allowed to drive.

Mrs. Lightner was mad. She got even madder when she was told, "That's the way the system works." That's when Candy Lightner decided to fight the system. She formed a group called Mothers Against Drunk Driving, or MADD, to work for tougher drinking and driving laws.

There were only a few members at first. "It's an uphill battle," Candy's friends said. "You can't win." But Candy Lightner won! Since 1981, in places all across North America, new and tougher laws are cracking down on people who drink and drive. MADD now has more than a million members and supporters. And Cari's story will never be forgotten.

Tips for Kids on the Road

- **Don't** get in a car with someone who has been drinking. Instead:

    -- Get someone else to drive you.

    -- Call a taxi or take a bus or subway.

    -- Call your parents or another adult to come for you.

    -- Walk.

- **Don't** let the people you know drink and drive. Take away their car keys if you have to.

- **Call the police** if you see a driver who looks or acts drunk.

- **Don't** stay at a party where people under 21 are drinking.

- **Think.** You might just save a life. And it just might be your own.

Drinking and Driving

It is against the law for anyone under the age of 21 to drink alcohol. It is also against the law for people to drive if they have had too much alcohol to drink. If a police officer thinks someone is not driving safely, the officer will want to know how much the driver has had to drink. But how can the police officer find out?

There is a test that shows exactly how much alcohol is in a person's blood. From even a small drop of blood, a police officer can measure a person's blood alcohol content, or BAC. If the BAC level is too high, the driver will be charged with a crime: it is called driving while intoxicated, or DWI.

There is also a way to tell how much people have had to drink by testing their breath. They breathe into a machine called a breathalyzer. It shows the amount of alcohol in a person's body.

These tests have made it easier for the police to stop drunk drivers. But the police cannot stop them all. There are still people who drink and drive. And there are still many people being hurt or killed in drinking and driving car crashes.

Alcohol and Illness

Most people know alcohol can cause headaches and vomiting. But alcohol can also hurt the heart, the liver, the kidneys, the stomach, and the brain. It can cause a loss of memory and can cause some kinds of cancer.

Because their bodies are still growing, young people are hurt by alcohol most of all. Dangerous chemicals like alcohol keep young bodies from growing the way they are meant to. These health dangers are not always felt right away. It may take a long time and many years of drinking to see the long-term effects of alcohol.

Alcohol can also hurt babies before they are even born. When a woman is pregnant, what she eats and drinks also feeds her baby. When the mother of an unborn baby drinks alcohol, the baby is given the drug, too. The alcohol in the mother's bloodstream is passed to her unborn child. And the baby can be born with physical or mental problems. No amount of alcohol is safe for a pregnant woman to drink.

One of the worst health problems with alcohol is that the more people drink, the more their bodies and minds get used to it. Getting used to alcohol is called tolerance. It means that people need two or three drinks to make them feel the way one drink did when they began to use alcohol. If they keep drinking, they may need even more alcohol to get the same feeling. Where does it stop? It may not stop until someone has a drinking problem. It can even lead to a disease called alcoholism.

Using alcohol changes the way the brain cells work so that people begin to need a regular dose of the drug to keep feeling normal. They soon develop a dependence on the drug. If they do not get a drink, they go through alcohol withdrawal. Withdrawal is the sick feeling drinkers have when they do not get alcohol. They may get headaches or feel nauseous. They may feel jumpy, confused, and nervous.

The symptoms of alcohol withdrawal may be different for different people, but one thing is sure to be the same: they will feel a craving for more alcohol. Some problem drinkers may drink only once or twice a day. Or they may drink only on weekends. They are called maintenance drinkers. Other problem drinkers make alcohol the focus of their lives. They drink at home and at work. They think about alcohol all the time. These people are addicted to alcohol. They are called alcoholics.

What is the difference between being drunk, having a drinking problem, and being an alcoholic?

- Being drunk means a person has more alcohol in his or her bloodstream than the liver can get rid of. In time, the liver will get rid of the alcohol, and the person will be sober.

- Having a drinking problem means a person is drinking too much and too often. Young problem drinkers may drink several times a week. They drink just to get drunk. Their drinking gets them in trouble with their families and friends, at school or at work, and sooner or later with the police.

- Being an alcoholic means a person has a very serious illness. Alcoholics can no longer control the need to drink because their bodies and minds are addicted to alcohol.

There are 10 million alcoholics in the United States and Canada. You might think of them as people who live in the streets or alleys, who wear ragged clothes, who beg for money to buy another drink. But that's not what most alcoholics are like. Most alcoholics live in nice houses, have families, and go to work. They are like most other people. But they have an illness, a serious one that affects both them and their families.

Alcoholism is sometimes called "the family disease." That's because the life of everyone—parent or child, husband or wife—is changed by having an alcoholic in the family. There are many problems to worry about:

- Will the alcoholic get sick?

- Will the alcoholic drink and drive?

- Will the alcoholic get fired from work?

- Will the alcoholic become violent?

- Will the alcoholic get drunk in public?

Children of alcoholics have a very hard time. They cannot count on their alcoholic parent to take care of them—or even to take them to the movies or a ball game. They learn the hard way that alcoholics have trouble remembering things: maybe a mother forgot about the class play, or a father forgot about the ballet recital. Many children of alcoholics are ashamed of the way their parents behave: maybe a mother was drunk when friends came over to play, or a father had too much to drink at a birthday party.

If you are a child of an alcoholic parent or parents, you might feel very lonely. But remember this: you are not alone. Many families are hurt by alcohol. And it helps to talk about your problems with other people you know and trust.

Some children of alcoholics blame themselves for the fact that their parents are problem drinkers. They might think, "If only I were a better kid, maybe my parents wouldn't drink so much." They think *they* are the reason their parents drink. *But the fact is they are not.* Their parents drink because they have a disease, not because of anything their children have ever said or done.

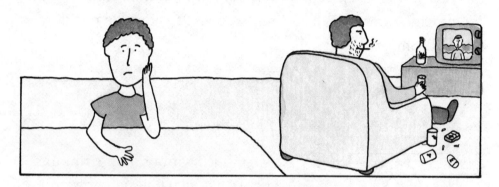

A person who is an alcoholic will always be an alcoholic. The need to drink won't go away by itself, and there is no permanent cure. But alcoholics can be helped. An alcoholic who stops drinking is called a recovering alcoholic.

It is not easy to stop drinking, and an alcoholic will need help. There are people especially trained to help alcoholics and their families. Doctors can help with health problems. And social workers, alcoholism counselors, and other people can help recovering alcoholics and their families lead happy lives again.

The "Not Me" Disease

Alcoholism has been called "a disease of denial." That means that many alcoholics deny the fact that they are problem drinkers. They refuse to believe that they are addicted to alcohol. But someone who *needs* to have a drink does have a serious illness. Denying the problem will only make it worse.

There are many ways to deny a drinking problem. Here are some of them:

"I'm not an alcoholic. I can stop any time I want."

Wrong. Alcoholics cannot stop drinking whenever they want. They have become dependent on alcohol to feel normal. So alcoholics need help to stop drinking.

"I'm not an alcoholic. I have a decent job."

Wrong. Most alcoholics do not look any different from other people. They are not just "bums." They may have good jobs and live in nice homes.

"I'm not an alcoholic. I'm only 13 years old."

Wrong. Alcoholics can be any age. One out of every 10 teenagers is an alcoholic. One out of every five teenagers has a drinking problem.

"I'm not an alcoholic. I only drink beer."

Wrong. An alcoholic does not just drink liquor. Most alcoholics drink only beer or wine.

"I'm not an alcoholic. I only drink at night."

Wrong. Alcoholics aren't drunk all the time. Many alcoholics drink only at a certain time of day.

"I'm not an alcoholic. Everyone drinks."

Wrong. The fact is that everyone does *not* drink. Many adults and most young people don't drink at all.

The Warning Signs
of a Drinking Problem

Do you know the warning signs of a drinking problem?

- drinking more than in the past
- hiding alcohol or sneaking drinks
- "blacking out" and forgetting things from drinking
- missing work or school from drinking too much
- lying about drinking
- driving while drunk
- needing a drink to have fun or to relax
- drinking alone
- drinking at odd times or in strange places
- thinking and talking about alcohol a lot
- drinking to help with problems
- having friends who drink

If you know someone who shows any of these signs, talk to an adult you know and trust. A person with a drinking problem needs help.

Alcohol and You

In the first chapter of this book, a young girl was invited to use alcohol. "Come on," her best friend said, "we're going to have a party!" Not everyone is asked to use alcohol in just this way. There are many ways young people are asked to try their first drink.

Will you be asked to drink alcohol? Will you have to make a decision about starting to drink alcohol?

You probably will, especially as you get older.

If you are asked to use alcohol, will you be able to say "No"? Here is something to think about: no one will be able to make that decision for you. It is a decision you will have to make by yourself, and *you* will have to live with your decision. It may not be easy to make the right decision about alcohol, but you can do it.

Remember that: *you can do it!*

Now that you have read this book, you know:

- what drinking does to your health and safety

- that drinking is against the law unless you are 21

- what drinking does to your family and friends

- how alcohol was used in the past

- how drinking can mess up your life

Saying "No" is not always an easy thing to do. Here are some ways of saying "No" that might make it easier for you:

- Think up an excuse: "I have to babysit."

- Change the subject: "Did you do the geography homework?"

- Be silly: "I'm hooked on ginger ale."

- Be surprised: "You're too smart for that!"

- Be disappointed: "What a dumb thing to do!"

- Think of something else to do: "Let's just go skating instead."

You can think of other ways to say "No." You can think of ways that are right for you. Practice them with your family and friends. Then, it will be easier to say "No" when you really need to.

It also helps to know that many adults and most kids do *not* drink. Here are the voices of some real kids who have thought about drinking—and decided to say "No":

"I don't like to be out of control."

"It tastes gross."

"It's against the law."

"My parents would kill me!"

"I want to have fun, not get sick."

"I need to be in good shape."

"I can have fun without it."

It may be hard to find a good reason to drink, but it is easy to find a good reason *not* to.

You can do it!

Glossary

addiction	the constant need or craving that makes people use drugs they know are harmful
alcohol	a drug found in beer, wine, and liquor
alcoholism	a disease in which a person can no longer control the need for alcohol
BAC	blood alcohol content: the amount of alcohol in a person's body
booze	another name for alcoholic drinks
breathalyzer	a machine that measures the amount of alcohol in a person's body
denial	when a person refuses to believe that a drug problem exists
dependence	the way the body and brain need a drug to avoid feeling sick
depressant	a kind of drug that slows down the way the brain works
drug	a substance that changes the way the brain works
drunk	another word for intoxicated

DWI	driving while intoxicated
ethyl alcohol	the kind of alcohol people drink
gateway drug	a drug that can lead to other drug problems
hangover	the sick feeling that people get after drug use
intoxicated	having more alcohol in the body than the liver can get rid of; also called drunk
isopropyl alcohol	the kind of alcohol used by doctors to keep things clean; also called rubbing alcohol
peer pressure	the feeling that you have to do what other people your age are doing
Prohibition	the movement to make alcohol products illegal
rubbing alcohol	another word for isopropyl alcohol
sober	not having alcohol in the body
temperance	the movement to slow down or stop people's drinking practices
tolerance	the way the body and brain need more and more of a drug to get the same effect
withdrawal	the sick feeling drug users get when they can't get the drugs they are dependent on

Index

Notes

Notes

Notes

Notes

Notes

Notes